John J Lundy, John J Lundy

Rinderpest

It's Prevention And Cure

John J Lundy, John J Lundy

Rinderpest

It's Prevention And Cure

ISBN/EAN: 9783741127212

Manufactured in Europe, USA, Canada, Australia, Japa

Cover: Foto ©berggeist007 / pixelio.de

Manufactured and distributed by brebook publishing software
(www.brebook.com)

John J Lundy, John J Lundy

Rinderpest

RINDERPEST,

ITS PREVENTION AND CURE;

AND

GYPSUM,

(OR SULPHATE OF LIME)

ITS ADVANTAGES AS A MANURE, DEODORISER, AND SANITARY AGENT.

BY

JOHN J. LUNDY,

F.C.S., F.R.S.S.A., &c., &c.

(Public Health Committee of the Town Council of Leith.)

EDINBURGH:
WILLIAM P. NIMMO, ST JAMES' SQUARE.
LONDON: SIMPKIN, MARSHALL & Co.
1865.

PREFACE.

The following remarks were, in the first instance, written as letters to a relative, and a dear friend of the author's, but who are unfortunately unacquainted with chemistry, as reasons for his having dared to advance to them the theory how to prevent, as well as to cure the Rinderpest and other contagious disorders to man and beast (now reduced to practice upon cattle), and as a collection of chemical facts, probably unknown to them, or which might have otherwise escaped their observation. As he went on, the matter increased to its present length. It was begun with a view to please those whom the author feels it a pride, a credit, and a duty to please and benefit, and in the words of a late fellow-townsman of his (Dr J. Alderson), who wrote an agricultural essay under similar circumstances as far back as 1796, he may say, that "should their publication excite the attention of any truly ingenious mind only till they produce refutation, or induce others by experiment to confirm them, the end of the author will be fully answered." Originally influenced by these motives only, a sense of duty to the public has since induced him to publish what he trusts will be found to be of use not only in itself, but to direct the minds of those more competent to grapple with the subjects treated upon. Probably the author's many imperfections will therefore be overlooked or pardoned, especially as the labour incurred by him has had to be done only by stealing odd moments and the "filings of time," and in hours which ought to have been devoted to repose.

JOHN J. LUNDY.

Leith, 2nd October 1865.

RINDERPEST,

ITS PREVENTION AND CURE;

AND

GYPSUM,

(OR SULPHATE OF LIME)

ITS ADVANTAGES AS A MANURE, DEODORISER, AND SANITARY AGENT.

AMONGST the many substances used as artificial manure, one of the most valuable to the Farmer (and most neglected by him), is Gypsum, or Sulphate of Lime, the value of which has never yet been properly appreciated,—a neglect arising from ignorance of its existence, or the proper mode of applying it to the various purposes for which it may be beneficially used.

Liebig says, in his Agricultural Chemistry, that the action of Gypsum chiefly consists in giving a fixed condition to the nitrogen or ammonia which is brought into the soil, by other manures or the rain, and which is indispensable for the proper growth of plants. The evident influence of Gypsum upon the growth of Cereals, Turnips, Potatoes, and Grasses, and the striking fertility and luxuriance of a meadow, upon which it is strewed, depends only on its fixing in the soil and retaining in the manure the ammonia, which would otherwise be volatilized. The Carbonate of Ammonia contained in rain water turns Gypsum, by double decomposition into Sulphate of Ammonia and Carbonate of Lime, and the Sulphate of Ammonia possessing no volatility is consequently retained in the soil, and is only given up as the plant requires it.

In some plants Gypsum does not form a component part, nor enter into immediate combination with them as food, yet it must strike every intelligent observer that the views of the Professor are just and correct, and we may fairly conclude that its effect on stable and fold-yard manure, guano, and rain water, is to combine with them, and render them more profitable by retaining their virtues, and consequently prolonging the duration of their beneficial effects until the growing plant assimilates them.

4

Sir Humphrey Davy, in speaking of those Crops which are found to contain Sulphate of Lime in any sensible proportion, (such as Clover, and Turnips,) says, it is probably combined as a necessary part of their fibre, so that it must originally be furnished by the soil. Hence the necessity for the Farmer applying it to those crops that require it, especially to land wherein Lime is deficient. Sulphate of Lime may thus be considered as having a two-fold action as a manure for all such crops, and a valuable aid to all other manures, less quantity of others being required, when it is used, to attain equal results.

As a manure to those plants in which it is found more particularly to exist, such as Clover and Turnips, it is better applied in the same proportions that are found to exist in the plant, as giving the best results, and affording a convincing proof to the Farmer of the benefits which Agriculture may derive from the assistance of the Chemist.

The action of Gypsum depends also upon its chemical affinity towards the other substances previously existing in the soil naturally or artificially. Possessing as it does the power of retaining and fixing the Nitrogen or Ammonia contained in the soil or manure, it thereby prevents the escape of that substance upon which their virtue mainly depends, and forms, by this new combination with them, Sulphate of Ammonia and Carbonate of Lime, the beneficial action of which is not confined alone to Clover and Turnips, but applies to all kinds of crops, acting as an efficient auxiliary to the culture of Barley, Wheat, or other cereals and leguminous plants.

For Clover it is recommended, even without the addition of other manures, as a top dressing. The produce of Clover, thus dressed, often doubles that which has had no Gypsum applied to it, as it prevents the land becoming Clover-sick, which may be often remedied by its application. Sir Humphrey Davy, on an analysis of Clover, calculated that an average Crop carries off, at least one bushel of Gypsum per acre ; no wonder, therefore, that soils containing so little of it should become tired of Clover or "Clover sick," the land being sooner or later exhausted and incapable of any longer supplying the demand the plant makes for it.

As a good crop of Clover often materially benefits the succeeding crop, it is recommended where the land has not got Gypsum naturally, that a dressing of from four to six cwt. per acre be applied broad-cast when the grass seeds are sown in April or May, and harrowed in, which is more beneficial than the ensuing Spring ; but if applied to seed, it is better used when the ground is moist, and adheres to the plant.

For Grasses and Pastures.—Sown broadcast during rain or when wet with dew, it will be found to increase the growth of white Clover, making it grow where it would not previously. Six to eight cwt. to the acre may be applied ;—but it will be found much better if ten cwt. be used. It will materially improve its growth, to use as a top dressing three cwt. to the acre, in combination with such other manures as contain Ammonia, and thus increase the effect of both. When it is known that sheep's wool contains 5 per cent. of Sulphur, it will in some measure indicate to the careful observer that the sheep must obtain it from the food, which must have been extracted from the land. It is therefore likely that unless it be supplied artificially to the soil, where it may not be derived from springs or other natural sources, these grasses which favour the growth of wool will gradually fail, and the weight and quality of the fleece diminish. If this Sulphur be extracted from the soil in the shape of Gypsum, then the plants upon which the sheep feed must take from the soil at least 13,000 tons of Gypsum per annum, reckoning the estimated number of sheep, ten years ago, in the United Kingdom, at 30 millions, and the average fleece at a little less than 4 lbs. only. It is highly probable that well gypsumed sheep farms would produce heavier fleeces of better quality, and with fewer vermin. It would be worth while for some of our extensive sheep farmers in hilly grass lands, with these facts before them, to try careful experiments, as beyond the use of sheep dips and smears (of the composition of which they are, as a rule, entirely ignorant), little has hitherto been done, except to grow the largest amount of food for sheep and manure for arable lands, without giving a thought to what food shall produce, not only the most flesh, but the largest quantity of the best wool as well as flesh and manure.

In Pastures where cattle are feeding, a slight dressing of two cwt. per acre will fix the Ammonia contained in their droppings, and sweeten the herbage for the cattle, preventing the waste of grass so often seen around them.

Notwithstanding the astounding results to land irrigated by town sewage in such localities as the Craigentinny Meadows (between Edinburgh, Leith, and Portobello), the production would be greater, and the herbage better and sweeter, were Gypsum sown broadcast over the land immediately after the sewage has been applied and soaked into the ground. It would render the locality more healthy, and remove all cause of complaint, provided the different conduits and channels were also arched over, or the service of the sewage supplied, as the health of the community requires it should be, by a proper system of piping. Indeed, it is a wonder that the local authorities of Portobello, dependent as it is for its prosperity on the salubrity of its atmosphere, should not have insisted upon the adoption of some plan for the deodorisation of the sewage applied there, or at all events that the channels by which it is conveyed be covered. It is equally surprising that the authorities of Leith should not have insisted upon the same plan being carried out, as, were it adopted, there would be nothing to prevent the land being gradually feued for villa residences from the west of Portobello all the way to Leith. This cannot be done so long as the meadows are irrigated by the present open sewers, as except the wind blows sea-ward, it carries its malarious influence over Portobello, Leith, or Edinburgh. Dr Littlejohn, in his recent able Report on the Sanitary State of Edinburgh, says distinctly of the Abbeyhill district (which is the nearest to the meadows) that "it is highly probable that a great part of the mortality of the Abbey, and some of the poorer districts of the Old Town, is in a great measure owing to the unhealthy breezes which blow so continuously during many months from the east over these meadows;" also, that "it has been plausibly conjectured that the special intensity and virulence of these winds depend on this contamination;" and further, that "it is difficult otherwise to account for the high death rate of the district of the Abbey, in which there is little overcrowding, and where only a small proportion of the population can be said to belong to the poorest classes."

For Barley, Oats, and Wheat, Gypsum combined with other ammoniacal manures will be of service, when applied alone to land, especially if applied as a top dressing immediately after sheep are taken from Clover or Turnips. It doubles the value of the manure they leave on the land; and for this purpose I would advise that they be confined in as small a portion of the field as possible at a time, by moveable fences, compelling them to finish one part of a field at once, and not giving too great space to stray over among uneaten turnips, as is commonly and unscientifically the present custom, which loses as it does the greater part of the ammoniacal properties of the manure they are placed there to deposit for succeeding crops of cereals, as well as wasting the Turnips not completely eaten. This, however, does not apply to Clover, as it is well known that sheep, if confined in too small a space for the number of them, destroy more than they eat. In America and in Germany it is universally used for these Crops. In Lincolnshire, Gypsum alone, sown and harrowed in with the seed, at the rate of eight bushels to the acre, showed a result decidedly superior to a Crop not gypsumed. I would advise Farmers to try the experiment, and I think they will be astonished at the result.

For Turnips, Gypsum is strongly advised, not to the exclusion of any of the many manures now introduced to the notice of the public, but in combination with them. It will be found to be an excellent auxiliary to their action, retaining their virtues for the benefit of the Crop, preventing the escape or volatilization of the Ammonia contained in them, and requiring only half the manure to be used to produce as good a crop.

In applying Gypsum to Turnips, if not previously mixed with the manure, I would advise the Farmer to sow five to six cwt. or more per acre on the land, previous to the application of other manure, and to harrow it in with it. No lime should be applied, as its action upon manure when put on the land immediately before or after the manure, has the contrary effect to Gypsum, by decomposing the Ammonia contained in it, and causing it to escape into the air.

Gypsum applied alone to Turnips acts as an excellent manure in lands where it is deficient. The *Mark Lane Express* says:—
" In our report of the Lincoln Agricultural Meeting, we incidently

noticed some Turnips exhibited by Mr Terry of New Holland, as a specimen of the extraordinary effects of Gypsum alone as a top-dressing to this important plant; we now give some further particulars,—the weight of one Turnip was 12 lb. 3 oz., and measured 30 inches in circumference: another was nearly of the same size, and the remaining one fell short only three or four inches in girth: several highly satisfactory proofs have been offered in this part of the country of the efficacy of Gypsum, when very fine, as a top-dressing for Turnips." As a Preservation from Fly a top-dressing of three or four cwt. of Gypsum per acre applied to the young plants, during wet weather, will have the desired effect, and also tend to destroy grub at the same time.

The use of Gypsum has been found in Cumberland, Westmoreland, Lincolnshire, Leicestershire, and many other parts, to be attended with the most satisfactory results in the culture of Potato Crops; and no doubt the application of it may again restore the Potato to its original productiveness.

There are instances that have answered well, where Potatoes have been planted with Gypsum alone, by applying fifteen to twenty cwt. per acre; others have tried it, in the ordinary way, with farmyard manure, giving eight to ten cwt. per acre, and strewing it upon the manure before planting the Potatoes in the rows. Two and a half cwt., each of Gypsum and Guano per acre answered exceedingly well, and from a very narrow inspection it was not perceptible that any had failed; while of others planted alongside with farm-yard manure alone, at least one-third did not come up.

The more Gypsum used to Potatoes the greater is the production of the Crop, and those who have tried it have found them, when taken up, to be much more productive, *with a finer* and *cleaner* skin—a sufficient proof of the restoration of health to the plant.

Potatoes, when cut for seed, should have plenty of Gypsum thrown amongst them, and be turned well over with a shovel, so that the Gypsum may have an opportunity of healing up the wound, caused by cutting. It is now no longer questionable that Gypsum forms a part of the Potato as well as the Turnip, and should by all means be supplied with a liberal hand.

It is of the greatest importance for the purpose of fixing the Ammonia, and its use cannot be too strongly urged, not only as an additional value to other manure, but as a sanitary agent for sweetening the atmosphere, and preventing the flow of urine into the water ponds, so often near to fold-yards, which not only lower the standard of health of the cattle, but often gives rise to disease, especially of a typhoid character.

The loss in farm-yard manure has been estimated at twenty per cent, or nearly one-fifth of the whole solid matter which it originally contains and gives the Crop a less benefit than one-half of the weight will do when the ammonia has been retained and fixed by Gypsum being occasionally strewed over the fold yard.

The practical Farmer who uses every effort to collect and preserve the manure which is put within his reach, is deserving of praise when he expends his money in the purchase of manures brought from a distance, of whatever kind they may be; but he, on the other hand, is open to censure, who puts forward the purchase of foreign and artificial manures as a substitute for those which are running to waste around him. Let every Stock Farmer, with the help of these facts, make a fair calculation of what is lost to himself and his country by the hitherto unheeded waste of the urine of his Cattle, and he will be able clearly to appreciate the importance of taking some steps for preserving it in future.

The urine of the Cow is not lost sight of in Holland. It is said to contain less water than that of man; but much depends on the food which has been eaten. A healthy average sized Cow gives on an average 1200 to 1500 gallons of Urine per annum, estimated by Professor Johnston to yield about 1,000 lb. of solid matter, of the annual value of from £4 to £5. This urine when six weeks old loses five-sixths of its Ammonia, unless it be fixed by Gypsum or Sulphuric Acid. To saturate the whole of the ammonia capable of being formed per annum by the urine of a single Cow of average size, would require about 700 lb. of Sulphuric Acid. It has been found that 750 gallons treated with only 15 lb. of Acid, produced equally as much hay as two and a-half cwt. of Guano, or one and a-half cwt. of Nitrate of Soda. It were however, far safer and handier to use Gypsum than Acid, as the Gypsum can be strewed in the byre daily, where Acid could not be used, and so prevent the loss of ammonia there, as well as in the tank. It also prevents the necessity of adding water to the urine in the tanks, which is done by some in at least equal quantity, as its dilution prevents a considerable loss of Ammonia during its fermentation, which dilution however, unnecessarily increases its bulk and does not effectually retain the ammonia like Sulphuric Acid or Gypsum. If there are, say eight millions of cattle in existence in the United Kingdom, the urine of which is worth only £4, its loss is equal to £32,000,000, sterling. It is impossible to estimate how much of this yet runs to waste, but one-third of it will amount to more than the whole Property and Income-Tax paid in this country in 1864.

Gypsum is a valuable agent to assist at comparatively little expense, in fixing the Ammonia in the dung and urine of the Cattle, and preventing its loss. I would recommend a small quantity being strewed upon the floors of the stables, cow byres, and fold yards, whereby it will not only promote the health and comfort of the animals, but of those who attend them, purifying the atmosphere they breathe, and preventing both from inhaling the ammoniacal vapour generated from the manure, which is so much money lost to the Farmer. Gypsum should be strewn on the floor immediately before and after the removal of the old litter, or before the fresh straw is put on, thus keeping them always sweet. It should always be strewed on the stable and byre floors after being well swept out, and before the operation of Lime washing, in order to retain the Ammonia liberated by the Lime wash falling upon the Ammoniacal deposits on the pavement and its interstices, in which case it were better that the Gypsum be damped with a little diluted Sulphuric Acid, to prevent the continuous action of the Lime. If used in sheep pens the ground would be drier, and tend to prevent foot rot, as well as double the value of the manure. The collection of the liquid manure, and drainage from the fold-yard is now becoming more general than heretofore, but it should be kept deodorised by Gypsum, not only to detain the Ammonia, but to keep the air surrounding the tanks sweet. The difficulty of applying liquid manure sometimes appears insuperable, but they may be obviated by adding Gypsum to it, and thus preventing the necessity of carting it upon a young growing Crop. Upon many light soils the damage done by carting liquid manure on the growing Crops is not to be compensated by the benefit the manure affords; and, in other instances, we find the want of the necessary apparatus of pump, cart, &c., is a bar to its adoption by many small Farmers.

One cwt. can be added to, say from seven to nine gallons of liquid manure, or the liquid from the tank may be added to the Gypsum on the ground, and mixed like mortar, adding more Gypsum if necessary until it assumes the consistency of Guano. Were all the Gypsum thrown at once into the tank, it would be difficult to mix it, without well stirring, therefore a small quantity should be put in daily, or dissolved with liquid manure or water to the consistency of paste before being put in the tank, when it could easily be stirred, and would not sink to the bottom, which it might do unless care be taken to prevent it, if the Gypsum be not properly manufactured.

If mixed on the ground, and after the Gypsum has absorbed as much of the fluid as it can do, it may be left to dry, and then fresh additions of the liquid may be made to it. The Gypsum will retain all the Ammonia and the valuable parts of the liquid, water only evaporating each time as it dries. It may then be kept until required for use like Guano, and will be found a most excellent manure for any crop desired. For every six cwt. of Gypsum used, forty-eight gallons of the very essence of farm-yard manure is preserved at a trifling cost, which would otherwise run to waste, fouling the water the cattle drink, and producing typhoid and other diseases to men and animals. It converts it by this simple plan, into a profitable manure of undoubted efficacy, in a less bulk and easier applied to the land than by the water cart, whether for broad-cast top-dressing, harrowing, or ploughing into the land; and especially when drilling every kind of seed, or planting Potatoes.

From railway stations and other public urinals a large revenue might be made if proper scientific means were taken, *without* creating any nuisance so great as now exists. Human urine is a fertilising manure of the richest quality. One hundred gallons, according to Professor Johnston, contain 68 lbs. of dry matter, worth at least six shillings,—the national waste in human urine alone being about six shillings per head per annum, and the urine of six men being equal to five tons of farm-yard manure, or enough to keep an acre of ground in good heart. Some years ago, in the " *Report of the Committee on Metropolitan Sewage,"* Mr Smith, of Deanston, said he considered the urine of *two* men sufficient manuring for an acre of land, and if mixed with ashes it would produce a good crop of turnips. By adding Gypsum when human or pig's urine is putrifying, its Sulphuric Acid will fix the Ammonia, and the Lime be converted into a Carbonate, which helps to dry up the Phosphates (present in human and pig's urine, although not in that of the cow, horse, or sheep) along with a portion of the animal matter. If it be so treated, and then dried, and the water evaporated by air or other gentle means, it can be again wetted with putrifying urine until perfectly saturated. Such saturation will fix more Ammonia, dry up more Phosphates, and also detain all the saline and other substances contained in it, such as Urea and other organic matters containing Nitrogen, as well as the Magnesia, Sulphate of Soda, Salt, &c.,—water only evaporating from it without any smell whatever. This plan is better than that of Dr Stenhouse, who, precipitating the Phosphoric Acid by Milk of Lime as Phosphate of Lime, found the produce when dried at 212° F., to contain 39 per cent. of Phosphoric Acid and organic matter, but only one per cent. of Ammonia. Or the Ammonia and Phosphoric Acid may be fixed as soon as putrifaction commences by a solution of Sulphate or Muriate of Magnesia by which means the double Phosphate of Magnesia and Ammonia falls to the bottom, and of which salt 7 lbs. may be obtained from 100 lbs. of human urine. Gypsum is by far the cheapest and most effectual, making a better manure, and if the Calcined Gypsum is used, and the mixture kept well stirred, to prevent its setting hard, it will take up about 33 per cent. of its own weight of water, when there will be less risk in loss of Ammonia in drying by heat, as well as the expense if not dried by air ; but should it set hard, it would be quite as good a manure when properly ground or powdored, and easier applied to the land, whether for " scaling " like Guano, or for drilling seed.

Pig's urine, as stated before, contains a considerable proportion of Phosphoric Acid (which is absent in that of the cow, horse, and sheep), adding considerably to its value as a manure when properly treated in the way I recommend for human urine, and it is satisfactory to know that in farm-steads all these can be allowed to drain into one tank. The whole of the manurial properties of the urine of man and domestic animals, although differing in their chemical constituents, can be thus preserved by one uniform method ; moreover, Gypsum, unlike Sulphuric Acid, can be applied by the most ignorant labouring man without endangering the person, the animals, or the crops.

Piggeries in large towns should never be allowed in crowded localities ; but wherever they are permitted they should only be where open courts are provided in front of the huts, and the courts should be frequently deodorised by Gypsum, whereby the animal, as well as its attendant, would escape breathing the foul stench otherwise arising, which must render the animal more liable to that disease from which few pigs are free in hot weather.

Mixtures of otherwise lost manurial matter, such as the urine of men, horses, cows, and pigs, which vary so much in their chemical proportions, will not like what has been proposed by some chemists as a rotation of manures, rendering it possible to grow more continuous crops of the same kind, through one compost made of all the liquid and solid animal and vegetable refuse of the farmstead with Gypsum. No single crop could be continued on the same land unless there be present naturally or artificially all the constituents of its composition, which possibly might always be the case without danger of exhaustion to the ground were it manured in a scientific manner by practical chemists, having previous knowledge of what was superabundant and all but inexhaustible in the soil by the crop proposed to be grown. I consider the farmer should be able to grow one kind of crop as continuously on the same soil by means of chemistry, or rotation of manures, as the manufacturer can continue to make one article in his factory or workshops. The urines, as well as the solid excrement of each animal, differ materially. For instance, the Phosphates found in the urine of the man and pig are not found in those of the cow, but go away from her in the milk to give bone-forming matter to the calf. It is possible that the urine of the bull or ox may contain Phosphates, and that of the mare and the ewe may contain none, for it is more than probable Nature lets it go into the milk, as it certainly does into that of the cow, but I cannot find anywhere that chemists have experimented for Phosphates on the urine of the bull, mare, or ewe. It is equally possible, and very probable, that Phosphates may be present in the urine of the cow, as well as the mare and ewe when not giving milk, as no authority I have consulted contradicts the idea of their existence, and I have neither had time nor opportunity to investigate it for myself; but if Phosphates are found under such circumstances, then the value of the urine of a farmstead is all the more valuable. One thing is certain, that urines contain many important saline properties not present in the solid excretions, and it may be, if all these matters were saved for use in a proper manner, and added to the vegetal refuse and solid excrement, they may be found, *as a whole*, to contain every necessary ingredient sufficient to produce continuous crops of any kind suitable to the land and climate. .

Dry organic manures, such as crushed bones, horn waste, woollen rags, shoddy refuse, or hair, unless changed in their composition before application to the soil decay slowly, and hence are not so quick in being beneficial as sea-ware, vegetal refuse, and animal excreta or chemically prepared manures. Hence it is found useful to apply such dry substances, where procurable, to Turnips; for as the other manures hasten the early growth of the plant, so these dry matters prolong its growth in the autumn months. But this double effect is produced equally well by using Gypsum to any manure that contains Ammonia, because the Sulphate of Ammonia and Carbonate of Lime then remains in the soil until taken up by the plant, or is washed away by the rains. Unless, therefore, this combined compost contain all the chemical constituents of all the crops a farmer grows in his rotation of crops, a *rotation of manures* should be introduced as a necessity where " high-farming " is aimed at.

It may not be inopportune here to insert Johnstone's table of the relative values of different natural manures in regard to their quantities of Nitrogen, being the quantity of each required to produce the same perceptible result as 100 lbs. of farm-yard manure :—

Farm-yard manure,	.	.	100 lbs.
Solid excrement of cow,	.	.	125 ,,
Liquid do. do.	.	.	91 ,,
Mixed do. do.	.	.	98 ,,
Solid excrement of the horse,	.	.	73 ,,
Liquid do. do.	.	.	16 ,,
Mixed do. do.	.	.	54 ,,
Mixed do. of the pig,	.	.	64 ,,
Mixed do. of the sheep,	.	.	38 ,,
Liquid blood,	.	.	15 ,,
Pigeon's dung,	.	.	5 ,,
Dry blood,	.	.	4 ,,
Dry flesh,	.	.	3 ,,
Feathers,	.	.	3 ,,
Cow hair,	.	.	3 ,,
Horn shavings,	.	.	3 ,,
Dry woollen rags, only	.	.	2½ ,,

Liquid blood, pigeon's dung, dry blood, and flesh, were better mixed with twenty times their weight of Gypsum or ordinary soil, to prevent their overaction upon the plant in contact with them, as well as to retain their virtues a longer time.

In Glue Yards and Tanneries the refuse heaps are (especially in summer weather) a constant source of annoyance alike to the manufacturers, their workmen, and their neighbours, through their fœtid smell, which can be altogether avoided by adopting the same plan. Were those heaps covered with a thin layer, and each succeeding addition sprinkled over with a little Gypsum, the waste matter would retain more Ammonia, and become a more valuable as well as a more reliable manure to the agriculturist, or the same result would be attained by mixing with each addition a quantity of weak Sulphuric Acid, sufficient to convert the Lime (which assists the Ammonia to escape) into Sulphate of Lime, which would detain it, and so increase its value and efficacy. To treat this waste with *Calcined* Gypsum, however, not only immediately deodorises it, but dries it sufficiently to enable it to be ground into a powder, from which the numerous nails can be sifted, and sold as old iron, which is of far more value than the refuse, or the cost of obtaining it.

The proprietors of Stable and Cow Byre dung middens in crowded localities should be compelled in like manner to keep them deodorised, as well when in a state of rest as during the removal of the manure, but specially when it consists of old rotten stuff, and also to deodorise the bottom of the midden immediately after it is emptied. Most of these places have au outlet to, or are close to the foot pavement, and the stench arising from them is in some places exceedingly prejudicial to the health of those who reside or have to pass near them. I have seen some stables so overcrowded with horses and cows, in which the ventilation was so bad, and the ammonia so pungent, that I could scarcely enter them without coughing violently, and could not remain in them without gradually accustoming myself to it. Seeing that Ammonia is the vehicle by which all gaseous contagions are carried about, there is no wonder that such stables and byres should become the hot-beds of such contagious disorders as influenza, pleuropneumonia, rinderpest, and epizootic aptha, not only generating but propagating some of these scourges to owners of cattle. Gypsum properly applied to such places would not only keep them free from Ammonia, but prevent the nitrification of the buildings, which so much damages them.

Graveyards and Cemeteries, especially in populous localities, but more necessarily at times when epidemic and epizootic diseases prevail, should be freely gypsumed to fix the Ammonia arising therefrom. This is the more imperative where the soil is sandy and open, as is the case in the churchyards of Leith and its neighbourhood, where it was stated, when an endeavour was made to close the churchyards of that place, *as an excuse for not doing so*, that total decomposition took place in less than seven years. If the ground, therefore, is so porous as to admit of such rapid decay, it would as surely permit the free upward escape of the Ammonia evolved as it admitted the Oxygen downward to assist the putrifaction to go on so rapidly. Hence the argument was turned against the propounders of it, and happily, those at one time pestilential places were virtually closed by raising the burial fees above those of the suburban cemeteries.

In a letter to the *Leith Burghs Pilot* of 21st September last, I stated, that in attempting to account for the origin of Rinderpest, it may not be inappropriate to notice that Sprengel says that in 100,000 parts of the urine of the cow, in a fresh state, ranging in a rich pasture, he found one part of oxide of manganese; and it were well to investigate how that one part of manganese gets into the system. If in the field in its normal state of feeding, then it may be that by the unnatural mode of stall-feeding upon draff, sewage grass, &c., it is deprived of food which contains manganese, or the atmosphere of the byres loaded with ammonia that it is not able to assimilate from its food enough to keep free from disease *when gaseous contagious matters are in the atmosphere*. If such be the case, it may be that the oxide of manganese, separated from the permanganate of an alkali (like any of Condy's Fluids) by the liberation of the ozonic-oxygen (the natural scavenger for all organic putridities), would be a cure for the disease by simply supplying the want of manganese in the system. To the healthy as well as diseased, manganese acts as a tonic like iron, and a portion of the liberated oxygen from the permanganate no doubt combines with the hydrogeneous gases, and passes off as water, precipitating sulphur, or forming sulphate of

14

potash or soda, in either case assisting to liberate the stomach, which
might otherwise remain charged with "wind" and undigested food,
that being a great desideratum in attempting to cure the disease.
Since then black oxide of manganese has been tried in Edinburgh
with varied success.

The Rinderpest Sanatorium officials of Edinburgh should, for the
sake of the health of the people of Leith, be compelled to apply
Gypsum to the burial ground. At present the animals are covered
with quick lime, and then simply covered by the soil. This is done
with an *erroneous* view to protect the health of the public. They
would be better buried without the lime, as the decay and consequent
evolution of Ammonia would be slower; or if lime be put on them,
which hastens their decomposition and production of Ammonia, they
should, after a slight covering of soil, have a thin layer of Gypsum before being finally covered up. Fields of battle and overcrowded graveyards are known fruitful sources of disease and pesti-
lence, and we may expect, if the Rinderpest be not soon stopped, that
this *so-called* Sanatorium may prove an awful scourge to the people
of Leith, placed as it is in such close proximity to the town. Rinderpest, Typhus, and all other gaseous contagions are no doubt
propagated by excess of Ammonia in the atmosphere. It may be
that Ammonia gives vitality to the virus, or intensifies it, as well
as being the vehicle of its propagation, carrying it about in the air
as surely as the down carries the seeds of the dandelion or other
weeds. Rinderpest may therefore be prevented, or at all events
mitigated with Gypsum, by which the farmer not only obtains the
benefit derivable from Gypsum itself, but also the economy due to
its improvement of the quality of all other organic manures.

Rinderpest may thus, by prevention alone, be at once swept
away through destroying the vehicle by which it is carried from
place to place. *Ronbiquet* says that gaseous contagious matters are
generally accompanied by Ammonia, which may be considered in
many cases as *the means through which the contagious matter receives
a gaseous form*, just as it is the means of causing the smell of innu-
merable substances of little volatility, and many of which have no
odour. *Liebig*, in 1840, said—" Ammonia is very generally produced in cases of disease. It is *always emitted in those in which
contagion is generated*, and is *always* present in the chambers of
those afflicted with a *contagious* disease." Nothing, therefore, can
be plainer to the practical farmer, than that chemical, if not medical
or veterinary science, has shewn how to prevent this scourge, although
the means may never have been before proposed. I stated in a
letter to the *Leith Burghs Pilot* on the 4th August, and the *Scots-
man* of the 15th inst., it can be done by the use of Condy's Red
Permanganate. It has been tried and proved effectual with cows
in the Rinderpest (which veterinary doctors say is a typhoid fever
analogous to that of the human subject, though not infectious
except to the bovine species); and on the 19th ultimo, Dr W.
Brewer (a London Physician of eminence), in writing to Mr H.
B. Condy, the Patentee, the only person who has hitherto manu-

factured the permanganates as an article of commerce, says :—
" The Committee of the National Association for the Prevention
of Cattle Diseases have deputed me to write to you, to inquire
if you can again supply them with the red fluid in twenty-ounce
bottles, as they have used up in experiments all they had. This
weak six cows, suffering from the Rinderpest, have been treated
with ounce doses of the red fluid in a quart of water three times
a-day, and !*to our surprise* three have recovered. Yesterday they
gave milk again, and this morning they were to all appearance quite
convalescent. This has rather taken us all by surprise. The Committee are recommending the internal use of the red fluid."

The partner of Mr Condy had a cow placed at his disposal, with
every symptom of Rinderpest, *from a herd of which many had died.*
After two days treatment she began to give milk again; the day
after gave five pints; the next day seven pints of milk, and
steadily improved. It has since been found by Dr Brewer, successful to give ounce doses of Condy's Crimson Fluid Permanganate
in *only* eight ounces, instead of a quart of water, acidulated with
sulphuric acid, as often as every hour, or even at less intervals.
After the first twelve hours or so it is desirable not to disturb
the cattle much at night. This fact, together with Mr Elliott's
experience, shows, that at first some causes of apparent failure were
due to using *too little* of the red fluid, and that little in too much
water, and a probability that the alcoholic stimulant with the
" hungering" system tried by some whilst the cattle are under
treatment is not so good, as giving the cattle some solid, or still
better, semi-solid nourishment with vegetal instead of alcoholic
stimulants, along with the administration of the oil and turpentine
or other medicine, with *strong* doses of Condy's Fluid, slightly
acidulated every hour. Warm, dry, deodorised, and well ventilated
stables are indispensable to speedy cures.

Mr Cyrus Alex. Elliott of Fulham, says in a letter to the *Times*
dated 27th September, " I purchased twelve bullocks at a small fair at
Findon, four miles from Worthing, Sussex, on the 14th inst. They
were kept in my field near there until Tuesday, the 20th inst., when
they were sent by rail in two trucks, which were said to have been
cleaned and limed, to New Wandsworth Station, where they were
fed with hay and watered, and driven through Wandsworth, over
Kew Bridge, to my farm in Boston-Lane, Hanwell. I saw them
every day. Late on Friday evening, the 22nd, one of them appeared
dejected and unwell ; but as it was dark, and the symptoms slight,
he was not treated until early the next morning, when he appeared
much worse—running from eyes and nose ; purged, discharge being
like soap-suds, and the stench from it intolerable. He was separated
from the rest, and half-a-pint of Oil of Turpentine and one pint of
linseed oil were given early. I saw him at 11 a.m. in the following
state : Ears and feet cold, breathing laborious, hair staring, &c. ;
repeated the turpentine and linseed oil, after which he seemed worse ;
soon, however, he was purged freely, and then appeared better.
Two wine-glassfuls of the solution of the permanganate of potash,

or Condy's Fluid, in a bottle of cold water, were given every hour. After taking the second dose he seemed warmer, and improved in every way. At six p.m. had little or no discharge, eyes looked better, breathing natural, urine passed freely, and he began to eat; he was so lively he could not be kept in the pen of hurdles. I considered him out of danger. On viewing the other bullocks I saw in five decided symptoms of the disease—yellow discharge from the eyes, a little from the nose, hanging down of the head, and an unmistakable dejected appearance. They were all—eleven in number—driven into a pen, and half-a-pint of turpentine and a pint of linseed oil given to each; they were then watered. The medicine acted both upon the bowels and kidneys. They then seemed better, had two wine-glassfuls of Condy's Fluid in a bottle of water given to them, and were fed with linseed tea and hay. These bullocks have all improved, and were pronounced yesterday by the inspector to be healthy. On Monday I saw the bullock first taken ill; he appeared to be sinking from exhaustion, and died about 3 p.m., when he was buried. After I left Hanwell on Saturday, the 23rd inst., I went to Worthing, and found my cow suffering from the disease; they had given the turpentine and oil twice during the day. This animal did not come in contact with the bullocks, though she was on the premises. When first taken ill she was purged, the motions being of a most nauseous smell. Symptoms at 11 p.m.: Suppression of milk, udder cold, ears cold, loss of appetite, and discharge from the eyes. In the morning seemed worse, the coat staring, and feet cold; gave two wine-glassfuls of Condy's Fluid, in water every hour, until she had taken a quart. After the second dose she seemed better, warmer, and gradually improved, and the next morning was convalescent; had gruel, fennel seed, and old ale given two or three times from Sunday to Monday night."

Owners of cattle should call in the assistance of veterinarians in preference to using quack nostrums, the composition of which is unknown, however many cures they may hear of by them, as it must be remembered, that cases occur less malignant than others, which recover under *all* kinds or *no* kind of treatment whatever, in the same way that men with good constitutions sometimes are able to fight out disease which kills others less robust.

Most of the cowfeeders in Leith have used and are using Condy's Fluid for disinfection, and Gypsum for deodorisation at my suggestion, *and up to the 28th of September there had not been* one case of Rinderpest in Leith, while scarcely any dairy in Edinburgh or *all round* Leith had been free from it. Since then one dairy has been attacked by Rinderpest, and lost by that scourge two cows. The proprietor had only used Gypsum and Condy's Fluid Permanganate *sparingly*, labouring under the impression he was doing enough, yet he had been foolishly giving his animals water to drink from a well within four yards of his midden, impregnated with the liquid filtering therefrom, although he had good Crawley water on the premises. The disease would no doubt have broken out sooner but for the precautions taken,—yet as the cowfeeders are getting a large advance upon the price of milk and cream, if they intend to appeal to the public to assist to pay for their losses, they should not spare the money obtained by the advanced price in adopting every possible plan for deodorising and disinfecting their byres, which ought to be kept as free from ammonia as their own bedrooms, instead of merely being content to let the disease take its chance. Drains ought to be all looked after, and proper stench traps at once applied, the byres white washed with Gypsum (*not quick lime*), and a plentiful supply every few hours, strowed thickly under their bedding, but specially in the water channels, and upon the droppings. We often hear of the cleanliness of our dairymen, but when closely looked into it generally applies only to the dairy and its utensils, whilst the byre is almost totally neglected, being with the yards and middens in a most disgraceful state,—arising more from want of knowledge, and careless indifference, than culpable negligence.

My opinion is, if the cows, *when first attacked*, are treated in a scientific manner, few, if any, will die. All healthy cattle, until the disease has left the country, should be given, say at least half-an-ounce once per diem of the Permanganate of an alkali (Condy's Fluid), and have occasionally a weak solution poured over them with a watering-can rose, or be spunged with it, and then rubbed dry with clean straw. I feel certain they will retain their health. Were it *universally* adopted to pour a small quantity into all the water given the cattle to drink, and the byres daily disinfected by Condy's Fluid, and kept constantly deodorised with Gypsum, to fix the Ammonia and sweeten the atmosphere, the Rinderpest would be swept from any locality in a few days, without resorting to the knife. If the knife be the only remedy, the study of veterinary science becomes a farce, and the attainment of progress in knowledge by our veterinarians completely stopped. Unless veterinary surgeons try to cure and prevent disease, of what use are they and their sanatoriums except to propagate it, which they cannot fail to do unless disinfected by some substance that will act perfectly, and *no other substance I know of will do it* without destroying their clothes except Condy's Fluids diluted with water. The cure I propose for typhus and other contagious diseases in the human subject, and for the Rinderpest in cows or sheep, is the same—namely, sufficient alcoholic stimulants and a good supply outwardly and inwardly of ozonic-oxygen, by means of permanganate of lime inwardly, and permanganate of soda outwardly by the warm bath, and for the purification of the atmosphere, whilst Condy's Crimson Fluid would do inwardly for all cases to the cows, and his Green Fluid would be cheaper for the stable and byre. In sick rooms the atmosphere may be deodorised and disinfected by any of these inodorous and non-poisonous fluids, diluted with water and exposed in any open vessel, or by very weak solutions poured out by a fine rose from a watering-pan; or, best of all in sick rooms, by one of the skilful little pneumatic glass perfume odorisers, which, by aid of the breath, throws out of a bottle an almost

B

imperceptible fine shower of spray. These glass odorisers can be used as well for deodorising a sick patient's bed-room, bed-curtains, and coverlet, as odorising a drawing-room or parlour with perfumes; or if fumigation by Chlorine be desired, the best plan is to put in a basin half-a-pint of Condy's Fluid, and add a wine-glassful of common Muriatic Acid, when a gentle stream of pure clean Chlorine will be slowly generated for many hours, resembling the odour of sea air and differing completely from that of Chloride of Lime, being due to Hypochlorous Acid, and not to Chlorine.

J. D. M. writing to the *Edinburgh Courant* on 30th Sept. says—"From analysis of the different parts of animals, and from experiments made in relation thereto, it would appear that a *disproportion* of oxygen to hydrogen, and of nitrogen to carbon, is the cause of rinderpest. If it be true that there is less oxygen and more carbon than is natural to the constitution of the various tissues, then the cure is simple. With fresh food and free exercise in open light and air, the double administration of active purgatives by the stomach, and of *pure oxygen* by the lungs, would in all likelihood be a cure. The cause is *within*, and no farther away than chemistry can reach. Let us take the advice of the *Times* of the 28th, and we shall soon know that rinderpest, cholera, and the like, have electro-chemical causes. Professor Faraday has proved that oxygen is magnetic. May not this element hold the balance of *powers* as well as proportions?" If this gentleman's experiments and ideas therefore are correct, there is no doubt that the "nascent ozone" of Mr Condy will supply the ozonic-oxygen required. It is natural to suppose that cows tied up for weeks together without any exercise, should, in the stifling ammoniacal atmosphere of the over-crowded dirty byres, be *unable* to inhale a sufficiency of oxygen for a healthy state. Indeed, I see no reason why cows should not in their turn all take exercise in the court-yards of byres where there are any, as a help to ward off Pneumonia, which very probably arises greatly from the atmosphere of a crowded byre, tainted as it is with ammonia, carbonic acid, and sulphuretted hydrogen, whilst the drains of the byre or yard, in many cases communicating direct with the sewers, being without any stench traps, add to the danger. It is found that after some of the cattle of a byre are dead, the remainder have the disease of a more benignant type and are easier cured, shewing that the overcrowded byre and the previous neglect of all sanitary arrangements (except keeping a clean dairy and dairy utensils), has been the pre-disposing cause of the attack, by lowering the standard of health of the animals. All medical men agree that overcrowding and bad sanitary arrangements predispose the human subject for typhus. It is equally so with cattle.

It has been found by practical experiment, that the dose for a cow may be at the commencement as much as one or two wine-glassfuls of Condy's Crimson Fluid, in as much water and slightly acidulated, every hour, by the mouth but *very soon* diminished in frequency, according to the judgment of the party in charge of the cattle; and if a veterinary surgeon use it, so much the better. An injection of a wine-glassful of the crimson fluid (without acid) in a quart of warm water, should be given as soon as attacked. Toddy, ale, or vegetable stimulants should be given at times after the fever is reduced, until no longer required, and all urine and excreta carefully disinfected by the permanganate or deodorised by gypsum, or both, or the treatment of turpentine and linseed oil recommended by Mr Elliott, and found successful in every case (except the first he tried) when accompanied by Condy's Fluid soon after their administration. His first failure arose from exhaustion, and it would no doubt have been a success had a vegetal stimulant or a dose of toddy or ale been promptly given. In the neighbourhood of Norwich, where sheep have been most extensively attacked with what the most experienced men declare similar to the Rinderpest, the treatment most approved is a gruel composed of Indian corn meal with oil cake, and a small dose of nitre and Epsom salts once a day, and the results ascertained by Dr Brewer were found very good.

The large manure heaps carted from the fold yards, and those abominable nuisances created by the Councils of Edinburgh and Leith, by the large accumulations of town fulzie placed in fields abutting on the high roads leading to the country, and at the railway stations, (especially that at Leith, which contains salmon and other fish offal and refuse from a large preserved provision manufactory, as well as fish and other refuse from Leith and the Newhaven district,) would be rendered almost harmless by the authorities compelling the contractors to apply a slight covering of Gypsum. This would add to the quantity, as well as improving the quality of the manure; and it is hoped the public will insist upon the suppression of this nuisance. In London and some other large towns night soil is sold as Animalised charcoal, being mixed with charcoal and Gypsum. In Paris, Berlin, and some other places on the Continent, it is mixed with Gypsum, dried, and sold as Poudrette, and said to be equal to thirty times as much horse or street manure, so that our Scotch fulzie could not be deteriorated but would be much improved by being mixed with or covered by Gypsum, which would fix the Ammonia as it is given off when it begins to putrify. If the other gigantic and pestilential nuisance at Craigentinny were abated, and efficient stink traps put to the sewer cesspools in the streets, house drains and waste pipes of the water cisterns (which latter too often communicate with the drains and sewers, damaging the water drunk and the air inhaled by the inhabitants), and the sewers ventilated by communication with the large gas works and other manufacturers' chimneys, where the poisonous gases could be burnt, rarified, or passed away above the houses, instead of often being ventilated by the rain water conductors, which generally just elevates the foul air to a level with and enters the upper and especially the garret storm-windows of houses, causing fever so often in the upper parts of houses, from which those who live in the lower flats and main-doors of the same building escape with impunity; together with the stoppage of sewage running into Lochend and the Water of Leith, the partial closing of the overcrowded church grave-yards, with the other improved sanitary measures already adopted in Leith, Edinburgh, and Portobello, it would make Leith and Portobello, if not Edinburgh, the most healthy populous towns in Scotland.

Gypsum, through ignorance in its mode of preparation, is sometimes found not to produce such results as are here stated, but when properly manufactured the results will be found truly wonderful to those who have not hitherto tried it. Indeed its use would as a rule save half the manure now used, and produce the same results.

Under the name of Alabaster, it is made into chimney-piece and other ornaments of numerous kinds, where the stone is found sufficiently compact to work with the chisel and lathe, as in Derbyshire. In a powder it is used to adulterate sugar confectionery, such as peppermint and other drops. After calcination it is used for the ceilings of rooms, and the finest kinds for the well known plaster images and castings. Indeed, this substance has such a number of uses in the arts and commerce, it would be almost endless to enumerate them: at any rate, it is a good friend to the farmer, though scarcely known by him.

———

Whilst the foregoing was in the press, a cow in Leith, labouring under a complicated case of Rinderpest, was placed under the care of Mr Thomas Todd, the Veterinary Surgeon here, who administered in the first stage a laxative, accompanied with a diuretic and sedatives, following them up by Condy's Fluid, slightly acidulated with Sulphuric Acid. As soon as the fever was reduced, the Fluid was continued, along with vegetal antiseptics, tonics, and astringents. On the evening of the third day she was much better, chewing her cud, and gave a trifle of milk. Next day she gave a quart, and same evening two quarts of milk, and was pronounced out of danger..

MOULD AND TOD, PRINTERS, EDINBURGH.

COOPER'S VETERINARY MEDICINE CHEST.

T. COOPER, Chemist and Druggist, 18, Walmgate, York, invites the special attention of Farmers, Graziers, and others, to the VETERINARY MEDICINE CHEST, which he has recently introduced, and which has received the commendations of many of his friends, who have tested its utility. He now offers it for general adoption, as being fitted with Medicines which have undergone a decided trial and proved *really efficacious*. Each Bottle and Packet is properly labelled, with ample directions for use. No Agriculturist should be without this Medicine Chest, which comprises every thing that may be necessary in case of emergency, where promptness of action is imperative for success.

The Chest is neat and compact, and the contents are of the very best Quality, and carefully prepared.

The Chest can be re-filled at any time, or any particular medicine supplied.

CONTENTS.

ALTERATIVE & CONDITION POWDERS.

Which are confidently recommended as invaluable for producing a beautifully smooth and glossy coat, keeping the horse free from humours, and in perfect condition; also curing those troublesome disorders—the Grease, Swelled Legs, &c.

THE NEVER FAILING GRIPE MIXTURE.

This Medicine never fails to give the animal immediate relief from the most excruciating pain. It would be advisable for every Farmer to have this mixture always by him, as the Gripes, on attacking a horse, soon terminates fatally, unless a timely remedy be administered.

WHITE OILS, for Bruises, Sprains, Swellings, Sore Throats, &c.
BLACK OILS, for Cuts, Bruises, Burns, Old and Fresh Wounds.
POTTAGE'S OILS, for Cuts, Bruises, Swellings, Kicks, Saddle Galls, &c.
GARGLE OILS, for the prevention of Gargle or Mortification in Ewes after Lambing, and for cuts, Green Wounds, &c.

TAYLOR'S VESICATING ESSENCE.

Decidedly the most valuable remedy that can possibly be used for Strains or Slips of the Joints and Sinews, Calloused Swellings of long standing, Splints, Ringbone, Wind Galls, Curbs, &c.; also for Inflammation of the Udder of the Cow or Ewe, commonly called "Garget" or downfall of the Udder.

COOPER'S ASTRINGENT MIXTURE, for Purging or Scouring in Calves & Lambs. This disease proving so often fatal, and consequently the cause of great loss to the farmer, the proprietor has been induced to make the value of this remedy more extensively known, feeling confident that if given in the early stage of the disease, it will not fail to effect a cure.

Also Castor Oil, Foot Rot Ointment, Cordial, Fellon, and Cleansing Drinks for Cattle

Price Complete, 21s.

The 25s. Chests contain in addition to the above—Cough Balls, Diuretic Balls, and Physic Balls. The 30s. Chests further contain The Yorkshire Specific, Red Mixture, and Restorative Condition Balls.

COOPER'S YORKSHIRE SPECIFIC.

This Specific will be found invaluable for the Undermentioned Diseases in HORSES.—It is a speedy, safe, effectual remedy for the Colic or Gripes, Gravel, &c., also for Mares after difficult Foaling.

CATTLE.—For Cows after Difficult Calving, Ulcerated and Bagged Udders. SHEEP, LAMBS, CALVES, and FOALS.—For Diarrhœa or Scour, Ewes after Difficult Lambing, and as an External Application for Fly Galls, Wounds, Swellings, Cuts, Bruises, Frosted Feet or Hands.

COOPER'S RED MIXTURE.

Is the safest and most effectual Remedy for Kicks, Sore Shoulders, and Backs occasioned by the collar or saddle; Sprains in the Sinews, Shoulders, Stifle and Whirlbone, Broken Knees, Windgalls, Bogspavins and all Flatulent Humours. It cleanses and heals all Wounds in the Joints and other parts where a discharge is not required; and as a generally useful remedy for external application, on all occasions.

COOPER'S RESTORATIVE CONDITION BALLS.

For Coughs, Colds, Fevers, Swelled Legs, Greasy Heels, &c., and for getting Horses into condition. These Balls are invaluable in cases of Loss of Appetite, Over-fatigue from Hunting or other severe exertion. They prevent Swelled Legs, assist Digestion, and produce a smooth appearance of the skin. Gentlemen who wish to have their Horses in good condition, should give one of those Balls twice or thrice a week, particularly at the spring and fall of the year.

In Boxes containing Six Balls each, 2s. 6d., or Seven Boxes for 15s.

ASPHALTE.

The advantage of the ASPHALTE FLOOR in Byres, Stables, &c., cannot be over-estimated on account of its Disinfecting qualities. Architects fully alive to the important laws of physical health now make free use of this valuable material in its application to dwelling houses : and the present crisis of a Cattle Disease loudly calls for its adoption.

ALEX. PORTEOUS,

Asphalter,

36 LEITH WALK,

EDINBURGH.

THE SCOTTISH FRIENDLY LIFE
ASSURANCE COMPANY.

Head Office—No. 5 ST ANDREW SQUARE, EDINBURGH.

BOARD OF DIRECTORS.

Chairman—DAVID THOM, Esq., Merchant, Leith.

HUGH ROSS, Esq., Merchant, Edinburgh.	THOMAS H. FERRIER, Esq., W.S
WILLIAM SCOTT ELLIOT of Arkleton, Esq.	WILLIAM PATERSON, Esq., British Linen Co.
ADAM BOLLAND of Gush, Esq.	THOMAS SPROT, Esq., W.S

Medical Officer—Dr SCONARBY JACKSON, F.R.C.P. & S.

Manager—FRED. H. CARVER, C.A. *Secretary*—JAMES T. SPENCE, C.A.

The "SCOTTISH FRIENDLY" offers to Assurers every advantage on the most moderate rates consistent with safety.

Annuities granted that contrast favourably with those offered by other Offices.

· Liberal Commission allowed to Agents.

Policies Indisputable after Three Years.

The highest surrender value given.

GRASS SEEDS,
For Permanent Pasture or Meadow.

The study of Natural Grasses has long engaged our attention, and we were among the first to recommend their use in the formation of PERMANENT PASTURE. After carefully noting the species composing the best Natural Pastures, as well as the results of combination in culture, under varied circumstances of soil, situation, &c., we feel confident that the agriculturalist will be a great gainer by using, in addition to the grasses and clover usually sown, an assortment of such of the Natural Grasses as are known to produce not only the greatest bulk of herbage, but also the largest amount of milking and feeding properties; varying those species, and the proportions of each, according to the nature of the land to be laid down. *When the selection is left to ourselves, it is requisite that, along with the order, we be informed of the nature of the soil, extent of the land to be sown, whether Statute, Irish, or Scotch Acre, and whether with or without a grain crop.*

W. DRUMMOND AND SONS, Seedsmen,
DUBLIN, and Stirling, SCOTLAND.

DRUMMOND BROTHERS,
SEEDSMEN, NURSERYMEN, AND FLORISTS,
52, GEORGE STREET, EDINBURGH.

DRUMMOND BROTHERS (Sons of Mr Peter Drummond of Stirling, N.B.) take this opportunity of intimating, that in September 1864, they commenced Business in Commodious Premises at the above address, where, by studied attention to the wishes of Customers, they respectfully hope to gain an annually increasing share of the confidence and trade of the Country.

Turnip and other **Agricultural Seeds** they use every precaution to obtain genuine and of pure STOCKS, and their **Mixtures** of **Clover** and **Grass Seeds** consist only of most suitable kinds.

With **Vegetable** and **Flower Seeds** they are equally particular, in procuring them of the finest quality and true to name.

The satisfactory results of the present season encourage them confidently to invite a trial.

SEED GRAIN AND VETCHES.
Fine Samples supplied at EDINBURGH MARKET Prices,—For Cash only.

NURSERY DEPARTMENT.

Having taken LAREFIELD NURSERY, FERRY ROAD, DRUMMOND BROTHERS are in a favourable position for executing Orders for every description of NURSERY PLANTS, especially **Hardy and Exotic Ferns, Roses, Rhododendrons,** and other **Ornamental Shrubs.**

RAILWAY AND STEAMBOAT COMMUNICATION
Afford facility of transit
TO EVERY PART OF THE UNITED KINGDOM.

Regarding FREE DELIVERY, see Catalogues.

Orders for **Exportation** executed with Care and Despatch.
September 1865.

DRESSER'S
"LIQUOR NICOTINE,"
OR
NON-POISONOUS SHEEP WASH.

A certain Specific for the Cure of Scab in Sheep and Lambs, also for destroying Tick, Lice, &c., and all Distempers on the Skins of Cattle and Dogs.

Price 2s. per Gallon, in Casks or Bottles.

PREPARED AND SOLD BY
R. DRESSER, Chemist,
PAVEMENT, YORK.

ILLUSTRATED
FARMER'S ALMANACK,
AND
Year-Book of Agriculture, Horticulture, and Arboriculture.

In announcing the publication of an Almanack in connection with the *Scottish Farmer*, the conductors believe that they will be able to secure for its pages not only a novelty in the arrangement, but a practical utility in the character of its information, which have not yet been realised in the literature of Agriculture. In place of giving a variety of information upon subjects which, however interesting as readable paragraphs, are at the best but of an ephemeral character, the conductors are of opinion that they will best serve the interests of that important and influential class to which their Almanack is addressed, if they make its contents of an abiding and permanent character—of information, in fact, of that class which can be made readily available in the varied pursuits of daily practice. If to this valuable character in the information given be added that of a careful classification of subjects by which this information can be readily got at or referred to, its value to the reader will obviously be enhanced. Having been for a long time under the consideration of the conductors, they believe that this essential desideratum will be obtained, so that a "commonplace-book," or "hand-book"—so to call it—of rural information, will be presented for the first time to the agricultural public, serving to them, it is hoped, the same practical office which well-known works have long served to the architect, the engineer, the builder, or general artisan.

A feature of the work will be a *résumé* of the labours of the year, and of the important improvements in practice, or of discoveries in theory, which have been made during its progress. The department of *Machines, Implements,* and *Farm Buildings,* has been intrusted to the care of Mr Robert Scott Burn, one of the authors of the "Book of Farm Implements and Machines," and of the "Book of Farm Buildings," and will receive that careful attention which its importance deserves.

Throughout the various sections of the work illustrations will be unsparingly used wherever these will serve to elucidate the text, and every pains taken to secure that completeness and comprehensiveness of treatment which will tend best to place the ILLUSTRATED FARMER'S ALMANACK in that position of practical utility which its conductors in its preparation are keeping so steadily in view, and which they are so desirous that it should occupy.

A limited number of **ADVERTISEMENTS** will be inserted, for which early Orders are requested.

SCALE OF CHARGES.

			£	s	d
One Page (displayed if required),	.	.	£8	8	0
Half a Page do.,	.	.	4	14	6
One Column do.,	.	.	8	8	0
Less than a Column,	.	per line	0	0	9
Bills of Eight Pages or under,	.	.	5	5	0
Bills Exceeding Eight Pages,	.	.	6	6	0

Size of Page same as the SCOTTISH FARMER Newspaper.

An Impression and Circulation of 10,000 Copies is guaranteed.

OFFICES: 26 Cockburn Street, EDINBURGH; and 150 Fleet Street, LONDON.